INSIDE
MODERN GENETICS

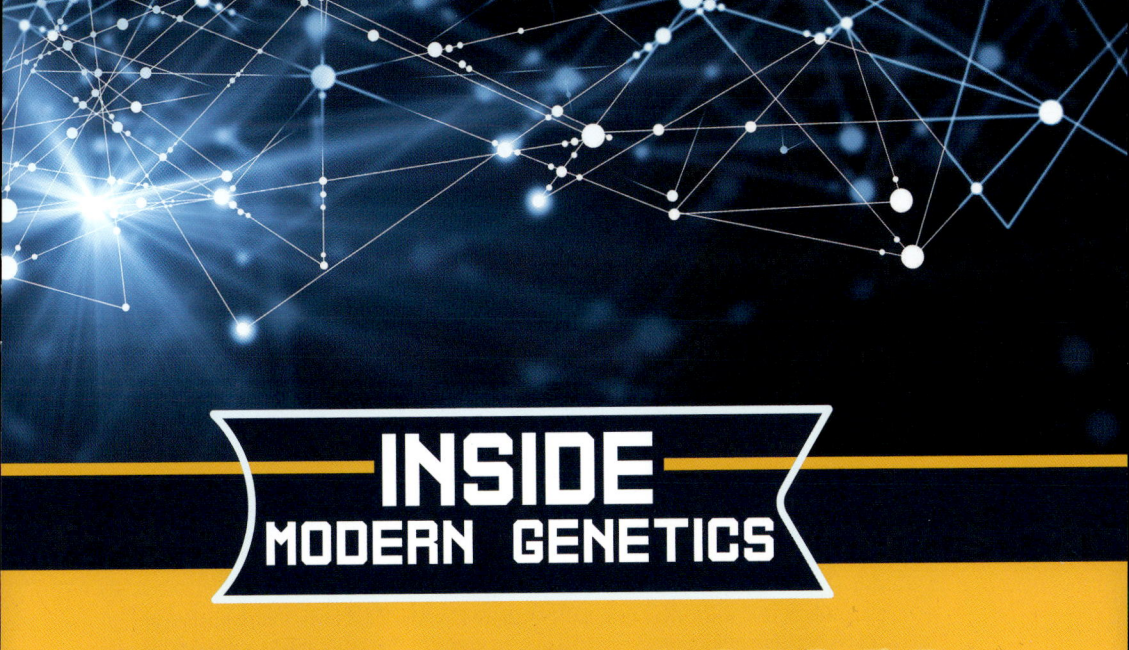

INSIDE BIOLOGICAL TAXONOMY

VERITY MILLER

Published in 2022 by The Rosen Publishing Group, Inc.
29 East 21st Street, New York, NY 10010

Copyright © 2022 by The Rosen Publishing Group, Inc.

First Edition

Portions of this work were originally authored by Mark J. Lewis and published as *Classification of Living Things*. All new material in this edition was authored by Verity Miller.

Designer: Rachel Rising
Editor: Jill Keppeler

All rights reserved. No part of this book may be reproduced in any form without permission in writing from the publisher, except by a reviewer.

Cataloging-in-Publication Data

Names: Miller, Verity.
Title: Inside biological taxonomy / Verity Miller.
Description: New York : Rosen Publishing, 2022. | Series: Inside modern genetics | Includes glossary and index.
Identifiers: ISBN 9781499470345 (pbk.) | ISBN 9781499470352 (library bound) | ISBN 9781499470369 (ebook)
Subjects: LCSH: Biology--Classification--Juvenile literature.
Classification: LCC QH83.M57 2022 | DDC 570.1'2--dc23

Some of the images in this book illustrate individuals who are models. The depictions do not imply actual situations or events.

Manufactured in the United States of America

CPSIA Compliance Information: Batch #CWRYA22. For further information, contact Rosen Publishing, New York, New York, at 1-800-237-9932.

Find us on

CONTENTS

Introduction 4

CHAPTER 1
Figuring Out Classification 6

CHAPTER 2
Linnaeus and a More Modern System ... 16

CHAPTER 3
A Matter of Evolution 36

CHAPTER 4
Clades, Cladists, and Cladistics 50

CHAPTER 5
The Future of Classification 56

Glossary 70
For More Information 72
For Further Reading 74
Index 76

INTRODUCTION

Say you're cleaning your room. There's a lot to go through, and you haven't cleaned it in a long time. Oops! There's a lot of stuff and a lot of chaos: clothes thrown on the floor, books stacked on your desk and bed, video games on the dresser with assorted other stuff. There's a sock hanging off your lamp, and your computer is nearly buried beneath school papers and books. Heaven only knows what's in your closet. (Certainly not your clothes.) Where can you possibly start?

It would probably be logical to start by sorting things out and organizing them. But how would you do that? You could sort by how long you've had the items. Your newest books and clothes could go in one dresser drawer and your oldest could go in another. But does that make sense? If you couldn't remember when you got something, you'd have a hard time finding it. You could also group things by how much you liked them: all your favorite books together with your favorite

Organization, whether it's of living creatures or of stuff in your room, can be complicated.

INTRODUCTION

clothes, games, and materials for your favorite school subject. That might be a little better, but it still probably wouldn't work that well. It makes more sense to group things by their function, or what they do: to put all your clothes together, all your books together, all your games together, and all your school supplies together. Then, when you need to find something, you'll know just where to find it.

This is also how scientists group living organisms. However, it can take far more research and study to know if a plant or animal really belongs with a certain group. An okapi looks a lot like a zebra, but it's far more closely related to a giraffe. Biological taxonomy is the science of classifying both living and extinct organisms. It's studying them to find out how they're different and how they're similar and putting them in groups to better study them.

Okapis are closely related to giraffes. The two species are the only living members of the family Giraffidae.

CHAPTER 1

FIGURING OUT CLASSIFICATION

Taxonomy isn't a new science. People have been trying to classify plants and animals for a very long time. The word itself comes from Greek: *taxis* means "arrangement" and *nomos* means "law." However, people had been classifying organisms long before the ancient Greeks did, especially for purposes of medicine and agriculture. The ancient Chinese and Egyptians both studied such things. In China, someone (tradition says Shennong, an emperor, though this is likely untrue) created a list of plants with medicinal value at least 3,000 years ago. In Egypt, records more than 3,000 years old also contain information on plants and their uses.

The word "taxonomy" has been used since the 1800s.

ARISTOTLE'S RULES

Starting with the ancient Greeks, we know a little more about efforts to classify plants and animals. Aristotle (384–322 BCE), a Greek philosopher and scientist, was the first person known to create a system to classify living things. In particular, he studied many forms of marine life while he was living on an island. Aristotle started by describing organisms carefully as the first step of classification.

Aristotle wrote about many things, including subjects as diverse as physics, poetry, and economics.

Think back to the example of your room. One way to classify items is by function. But before you can put an item with others of its own kind, you need to identify and describe the item. Describing the function of an item in your room is easy because you've probably used (or played or read or worn) it before. You probably even saw the package it came in and you have all that information.

Describing a newly discovered species is not so simple. There is no label that says "okapi" or "orca." Scientists must measure and study an example of the species and dissect it. They should also observe the new creature's or plant's habitat and its behaviors. Of course, people have been doing this for thousands of years, but Aristotle pointed out that all those observations are not enough.

An organism's traits or features must also be carefully compared with the traits of similar organisms. People used behavior, color, and habitat to classify animals and plants. But some classifications are more complicated. Killer whales (orcas) and great white sharks, for example, both live in the ocean. They are both swimmers. Each animal also has a dorsal fin that helps keep the animal upright in the water. Both creatures are also predators, which means they hunt and kill other animals for food. Are they closely related? Actually, no. Aristotle pointed out that only some characteristics were appropriate for use in classification.

Consider the following: Many kinds of animals have two eyes. Many aquatic animals

FIGURING OUT CLASSIFICATION 9

Great white sharks and killer whales may look slightly similar, but they're very different creatures.

(those that live in water) have fins. Many different kinds of animals are predators. These traits are not reliable for classification purposes, however. Aristotle worked to find more reliable traits to categorize living organisms. Does the animal lay eggs, like birds, fish, and reptiles do? Or does the animal give birth to live young, as mammals do? Is the animal cold-blooded, as reptiles and fish are, or warm-blooded, like mammals and birds? Does the animal breathe using gills or lungs? These traits proved to be far more reliable for classification. Consider the great white shark and killer whale. The great white shark produces eggs that it holds inside its body, while the killer whale gives birth to a well-developed baby whale. The great white has a body temperature that is the same as the ocean around it. The killer whale maintains a steady body temperature of about 98° Fahrenheit (37° Celsius). It has blubber to insulate itself from the cold ocean water.

The great white shark breathes using gill slits. The shark's forward movement forces water into its mouth, through the throat, and out through the gill slits. The shark uses its gill slits to extract oxygen from water flowing across its gills, so it never needs to surface as long as the shark keeps moving. The killer whale uses lungs to breathe, just as humans do. It must surface to take giant breaths. It can then hold its breath for as long as 15 minutes! Using Aristotle's early system, it becomes clear that although both whales and sharks swim and have similar color and shapes, they are

FIGURING OUT CLASSIFICATION

not as closely related as whales and people are—even though humans don't live in the water.

Naming new organisms has been an issue throughout the history of classification. People have described living organisms for thousands of years, but there was no common system of naming them. Often, they were just given names that described where they were found, like "flower of the river." The same organism might have 20 different names, which led to confusion. Aristotle put similar animals into groups called genera (plural for genus). These were general categories for any creatures or plants that Aristotle saw as being similar. Then he further divided the genera into different species based on further characteristics. This is called binomial definition. "Binomial" means "two names."

Great white sharks may look somewhat like whales, but they're not that closely related.

ALONG THE CHAIN

Aristotle put living things in order from lowest to highest on a sort of chain. Humans were at the top of this chain, which he thought of as perfect, fixed, and unchanging. The creatures and plants on the chain had no particular relation to each other, and the order wasn't supposed to be a mark of evolution. It simply went from the simplest plants at the bottom up through more complex organisms at the top.

This idea was progress in the field of classification, but it had a number of problems. There was no clear idea what it

Dogs and cats are both mammals, but which would be higher on Aristotle's chain?

DIVIDING THINGS UP

Aristotle first divided all organisms into two categories: animals and plants. From there, he divided animals into groups with red blood and without red blood, which tended to match the modern categories of vertebrates and invertebrates. Then he divided the animals into groups based on air, water, and land. He divided plants into small, medium, and large categories. Can you see the flaws in these groupings? For example, most birds fly, so they'd be considered to be in the air category with flying insects and bats. However, ostriches, along with a few other birds, can't fly. They'd be in a completely different category with land-based animals such as cats, although they're not related to them at all. What about ducks? They swim, fly, and spend time on land. Frogs and alligators live in the water and on land as well. Which category do they go in? There were problems with the plant category as well. Small plants aren't necessarily related to each other even if they're close to the same size. Aristotle admitted that the categories (which were also broken down further in some cases) did not fit every organism perfectly, but his system was an important start.

meant to progress along the chain. What did "higher on the chain" mean? Did it mean "more intelligent than"? Or "more powerful than"? Or "morally superior to"? And what is superior—an oak or a maple? A dog or a cat? Were they more or less equal, and if not, what made one higher than the other?

It was also an issue that the chain didn't mark the effects of evolution. Aristotle thought life on Earth was unchanging. Still, his work was important. He was the first scientist we know of who created a system for classifying living organisms, and the basics of his system were in use for more than a thousand years.

The earliest classifiers of life brought some order to the thousands of living things around them for two reasons. People tested and grouped plant and animal substances for use as medicine. Aristotle was very curious and wanted to figure out how things were related. Aristotle created the basis for a standard system of naming and showed that the classification of living things requires two steps: the thorough description of each organism and the thoughtful comparison of similar organisms. He was also the first to try to put all living organisms in order. His ideas influenced the way people looked at living things for many years. One of his students, Theophrastus, followed in his footsteps, categorizing and studying plants and affecting how we think of them today.

FIGURING OUT CLASSIFICATION

Would ducks be in the air, land, or water category? They spend time in all three environments.

CHAPTER 2

LINNAEUS AND A MORE MODERN SYSTEM

After Aristotle's time, however, a lot of his work was lost or forgotten in the West. The fall of Rome meant the loss of many works of writing and art. As the years went on, fewer people read ancient Greek, and Latin translations of works were rare. In addition, fewer people studied philosophy and such things at all.

Fortunately, this wasn't the case everywhere. By around the 10th century, in Baghdad and other places, Arab scholars were studying the writings of Aristotle and others, preserved and protected through the years. They translated the writings and built on them with their own scholarship. As traffic to the East grew, people from the West encountered the works of Aristotle again. They became the foundation of medieval philosophy, theology, and natural science.

AFTER ARISTOTLE

The Renaissance of the late 1500s, following the Middle Ages, brought more interest in studying and categorizing the natural world. Renaissance scholars often studied many different fields of research, and biology became a growing field. Many doctors were also botanists because plants were often used to treat

LINNAEUS AND A MORE MODERN SYSTEM

illnesses, and people began writing books with information grouping these plants and containing drawings of them. They also studied the human body in greater detail.

Physician Andreas Vesalius (1514–1564) studied biology and wrote a major work on the anatomy of the human body in 1543. Andrea Cesalpino (1519–1603) was a doctor and botanist living in Italy during the 16th century. Cesalpino studied Aristotle's approach to classifying organisms and used Aristotle's work to come up with his own classification method.

Andrea Cesalpino was one of the first to classify plants in a way that wasn't related to medicine.

Cesalpino is often called the first taxonomist because he used a logical approach to classification. Cesalpino carefully analyzed and compared all the plants he knew before selecting the characteristics he used to divide them into groups. Cesalpino classified more than 1,500 plants in his publications. He decided that seeds and seedlings were important characteristics to consider when dividing plant groups.

Following in Cesalpino's footsteps, two Swiss brothers, Gaspard and Jean Bauhin, studied plants in the late 1500s and early 1600s. In 1623, Gaspard Bauhin published a book that listed more than 6,000 species of plants. He grouped plants by genus and tried to keep the species names as short as possible. The book also helped clear up some of the confusion caused by repeated naming of plants by many different people over the years. Jean Bauhin published another work with thousands of very detailed descriptions of plants.

Toward the end of the 1600s and into the 1700s, interest in botany and zoology grew even more. One of the next significant scientists to work on classification was John Ray (1627–1705), an English scholar and naturalist who was also an ordained Anglican priest. Ray wrote many books on a wide range of subjects, including plants, mammals, fish, birds, insects, theology, and natural history, starting with *Catalogue of Cambridge Plants* in 1660. Ray wanted to find a classification for what he considered the divine order of creation, and he believed that he could better understand the wisdom of God by studying the natural world.

Most botanists focused on one or two characteristics when classifying plants. Ray looked at the total morphology, or the form and structure of the whole plant, and he influenced other botanists to do the same.

Ray also noticed that plants could seemingly be neatly divided into two large groups based on their number of cotyledons, leaflike parts of the embryo that serve as storage

LINNAEUS AND A MORE MODERN SYSTEM

John Ray was the first person to come up with a biological definition of "species."

> Ray used the terms "monocot" and "dicot" to classify plants.

organs for nutrients. If a plant's seed contained one cotyledon, Ray called that plant a "monocot." If a plant's seed contained two cotyledons, Ray called that plant a "dicot." This worked better than past systems that only looked at part of the plant. However, while monocots and dicots are still recognized as subclasses of plants in some more recent classification systems, the dicots are no longer recognized in systems based on evolutionary relationships.

Ray also looked at more details of the anatomy of the animal world, including lungs and hearts, and recognized the class of mammals. He also was one of the first scientists to promote the idea that fossils were once living organisms. He did not, however, believe that God would let living things go extinct. Instead, Ray thought that fossils that showed unfamiliar plants or animals were simply from organisms that hadn't been discovered yet, possibly because they lived in other parts of the worlds.

THE ADVENT OF LINNAEUS

As all this work by scientists with widely varying systems of classification was going on, the work of one Swedish botanist stood out. Carolus Linnaeus (also known as Carl von Linné) was born in 1707. He would forever change the ways people looked at the organization and naming of species and is considered the father of modern taxonomy.

Like many noted taxonomists, Linnaeus spent his childhood fascinated with plants—and with their names. His father, an amateur botanist, taught him the names of all the plants in their garden. When Linnaeus went to school, he often skipped class to look for plants in the countryside. Linnaeus's classmates nicknamed him "little botanist."

Carolus Linnaeus was born in this house in Småland, Sweden. He was the oldest of five siblings.

Linnaeus's parents wanted him to become a priest. However, Linnaeus showed little interest or talent for that calling. A doctor who taught at Linnaeus's high school was impressed by his knowledge of plants. He convinced Linnaeus's father that Linnaeus should study medicine so he would be able to use his knowledge of plants in his career. However, it took a lot of memorization at the time because most scientific plant names were long and in Latin. They often described plants by common parts that weren't necessarily distinctive. There were also many systems of organizations, not one used by most people.

In time, Linnaeus became a doctor and respected naturalist. He studied plants, animals, and minerals. He also thought a lot about how to best classify organisms. Linnaeus knew firsthand that the ways of naming plants and animals were confusing and cumbersome and that there was no overall system. He had ideas to change this.

BY TWO NAMES

Linnaeus started consistently using a reduced name for organisms: one name each with two parts. In 1753, he published a plant book that introduced his binomial nomenclature method of naming organisms. He was not the first person to give organisms two-part names, but he was the first to consistently use such a system. Linnaeus applied some of the genus names created by the Bauhin brothers for the first name, which is used to describe a whole group of similar organisms. The second name in Linnaeus's system was a one-word species name.

Sometimes Linnaeus took this word from an existing name. Sometimes he made up his own. Linnaeus was the first person to use species names that did not always describe the plant or animal. He sometimes named plants after people he knew. For example, he named a small weed *Siegesbeckia* after his

LINNAEUS AND A MORE MODERN SYSTEM

> Many people also consider Carolus Linnaeus one of the founders of modern ecology, or the science of the relationships between organisms and their environments.

rival and critic, botanist Johann Siegesbeck. Linnaeus's friend and fellow botanist Elias Tillandz disliked traveling by boat. Linnaeus named a genus of plants that does not grow well in damp soil *Tillandsia*, after his friend.

Many species named during Linnaeus's time still have the same name today. Scientists still use Linnaeus's binominal system to give newly discovered species a name. The genus name, together with the species name for an organism, is called its scientific name. As in Linnaeus's time, scientific names for new species are still created using Latin or Greek words.

Scientific names are always italicized. The species name often describes a main characteristic of the organism. For example, all maple trees belong to the genus *Acer*. *Acer* means "hard" or "sharp" in Latin, and it refers to the sharp points on the leaves of most maples. The scientific name for sugar maples is *Acer saccharum*. The sweet sap of sugar maples is used to make maple syrup. *Saccharum* means "sugar" or "sweet" in Latin.

The sap of the *Acer saccharum* is used to make maple syrup and candy.

LINNAEUS AND A MORE MODERN SYSTEM

> Lions are part of the same genus as tigers, leopards, snow leopards, and jaguars.

A lion is a member of the genus *Panthera*, and its scientific name is *Panthera leo*. Tigers are a member of the same genus, with their scientific name being *Panthera tigris*. Every member of this genus (which also includes the jaguar, leopard, and snow leopard) is one of the big cats, with certain cranial (skull) features in common.

A SYSTEM OF GROUPS

Linnaeus created a further system of organization in addition to his naming system. It was arranged in a hierarchical order. A hierarchy is an arrangement that divides large groups into smaller groups. The small groups can be placed in order beneath the big group.

For example, schools—high schools, in this example—have a hierarchical order. All students at the school are high school students. Within that group of students, the students are usually divided into seniors, juniors, sophomores, and freshmen. This follows an order of how long they've generally been at the school.

CAROLI LINNÆI
Equitis De Stella Polari,
Archiatri Regii, Med. & Botan. Profess. Upsal.;
Acad. Upsal. Holmens. Petropol. Berol. Imper.
Lond. Monspel. Tolos. Florent. Soc.

SYSTEMA NATURÆ
Per
REGNA TRIA NATURÆ,
Secundum
CLASSES, ORDINES, GENERA, SPECIES,

Linnaeus's *Systema Naturae* has many detailed drawings and descriptions of Earth's organisms.

The students can then be further divided into even smaller groups such as homerooms. Just like with Linnaeus's system, the smaller groups have to fit neatly into the larger groups for the hierarchy to make sense. One student (or organism, in the case of Linnaeus) cannot fit into more than one of the larger groups. For example, school clubs would not be a good example of groups to include in a hierarchical diagram of a high school. Club members would not all be in the same year of high school, and not all students may be in a club. Linnaeus's system was a neat hierarchical system.

When Linnaeus was in his late 20s, he published his system for classifying animals and plants in a book called *Systema Naturae* (*The System of Nature*). The original edition of *Systema Naturae* was simply an 11-page list of all discovered plants and animals Linnaeus knew of, classified according to his new hierarchical order. Throughout the years, Linnaeus added to his book. By its 10th edition in 1758, it was a huge two-volume publication.

From Aristotle's time, organisms had been grouped into the two very large kingdoms of plants and animals. Then they were grouped with similar organisms in much smaller groups on the genus level. Linnaeus created more levels in between.

He created seven descending groups in his hierarchy. Linnaeus's hierarchical ranking started with the largest group, kingdoms, and went down through phyla (plural for phylum), classes, orders, families, and genera (plural for genus) to the smallest unit of classification, species. *System Naturae* was the first classification scheme to group humans with monkeys as primates in the order Primata.

Linnaeus's colleagues recognized he had a talent for putting things in order. Unlike Aristotle, Linnaeus did not often run into the problem of finding an organism that fit into more than one of his big categories.

Linnaeus continued to add new species and findings to his *Systema Naturae* over the course of his career. He published 12 editions of the book before he died. Linnaeus once remarked to a friend that he was upset about all the work he was putting into it. He, like many scientists of his time, simply thought that most living creatures had already been discovered and named! Even today, however, it is widely believed that only a small fraction of the actual living creatures on Earth have even been discovered. As many as 90 percent of plants and animals are yet to be found and classified.

Linnaeus's hierarchical system allowed scientists of the time to group living organisms based on increasing morphological similarity. Newer classification systems kept Linnaeus's basic animal hierarchy but ignored his classification of plants by flowers. Plant taxonomists preferred John Ray's use of total plant morphology.

	SUGAR MAPLE	ROCKHOPPER PENGUIN	PAINTED LADY BUTTERFLY
KINGDOM	Plantae	Animalia	Animalia
PHYLUM	Tracheophyta	Chordata	Arthropoda
CLASS	Magnoliopsida	Aves	Insecta
ORDER	Sapindales	Sphenisciformes	Lepidoptera
FAMILY	Sapindaceae	Spheniscidae	Nymphalidae
GENUS	Acer	Eudyptes	Vanessa
SPECIES	Acer saccharum	Eudyptes chrysocome	Vanessa cardui

With the hierarchical nature of Linnaeus's classification, it became common to use the dichotomous key, another tool of classification. "Dichotomous" means divided into two parts. The dichotomous key has a series of questions that separate things into two groups based on the answers.

It's used for plant identification, among other things. Imagine a girl is hiking through a forest not far from New York. She comes upon an unfamiliar tree and wonders about it. Her dichotomous tree key asks a series of questions about the tree. Each question has only two possible answers. The first question asks, "Is the tree growing in the eastern or the western part of the United States?" She selects "eastern" and is directed to the first half of the book, where eastern trees are described. There, she is asked what kind of leaves the tree has. "Does it have cones and needlelike or scalelike leaves or thin, flat leaves?" She sees that the leaves are thin and flat and is directed to turn to a certain page. She is then asked if the leaves are simple or compound. They're simple, so she turns to another specified page. There, she's asked if the leaves are opposite each other on the same twig or alternating on different sides of the twig. The questions continue until she can identify the tree: a sugar maple.

MORE WAYS TO CLASSIFY

Scientists today do not always use morphology to classify organisms. However, morphological similarities can be used to explain how some organisms are classified. For example, it is easy to see that a sugar maple tree and a rockhopper penguin are both living organisms. They both grow, change, and reproduce. It is also easy to see that sugar maples and rockhopper penguins belong to different kingdoms. The sugar maple is a plant. Plants belong to the kingdom Plantae. A penguin is an animal, so it belongs in the kingdom Animalia. A painted lady butterfly is an animal too. Penguins and butterflies both have wings but are very different animals. Look at the chart on the opposite page to see how the classification of rockhopper penguins and painted lady butterflies differs.

INSIDE BIOLOGICAL TAXONOMY

Although rockhopper penguins and painted lady butterflies both have wings, they're very different creatures.

A rockhopper penguin is in the phylum Chordata. All animals in this phylum have three particular structures at some point in their life cycle: a notochord, a hollow dorsal nerve, and pharyngeal slits. A painted lady butterfly has a segmented body and jointed appendages, so it belongs to the phylum Arthropoda.

A rockhopper penguin has feathers. Like all birds, the rockhopper penguin belongs to the class of animals with feathers, called Aves. Painted lady butterflies have antennae and six pairs of legs, so they belong in the Insecta class.

Penguins are funny-looking waterbirds that stand upright. All penguins belong to the order Sphenisciformes and the family Spheniscidae. All butterflies have a long tube, called a proboscis, that they use to suck nectar. All butterflies belong to the order Lepidoptera. All butterflies that have short front legs that they can't use for walking belong to the Nymphalidae family.

Rockhopper penguins share the genus *Eudyptes* with some other species of penguins. They all have crests, or raised feathers on their head. All butterflies in the *Vanessa* genus have things in common too, including basic colors (orange, white, and black) and eyespot (ocelli) patterns.

Linnaeus and many scientists of his time classified organisms based only on physical characteristics or morphology. Since that time, however, scientists have discovered other tools for identifying and classifying organisms. Some study how different animals develop before they are born. The science of embryology has revealed relationships that are not obvious just from looking at animals. For example, the blue whale, which is a mammal, and the rattlesnake, which is a reptile, both have similar protective sacs that surround their embryos. The crocodile and the bald eagle both lay eggs with thick shells. Each case suggests that the organisms share a common ancestor.

Taxonomists have always looked for homologous characters. These are traits that two species share because they have a common ancestor. For example, humans and bats both have five digits at the end of their forelimbs. These digits form fingers in humans. In bats, the five digits form their wing bones. This is called homology.

Sometimes two species will look similar even though they are not closely related. They may have independently adapted to similar environments. Duck-billed platypuses lay eggs like birds and reptiles do. As its name suggests, the duck-billed platypus has a bill that resembles a duck bill. Platypuses also have hair, and mother platypuses produce milk to feed their babies. Biologists have decided that platypuses are mammals. They inherited their hair and milk-giving traits from a common ancestor they share with other mammals. The duck's bill and the platypus's bill are called homoplastic traits. "Homoplastic" means similar in structure but not in origin. Ducks and platypuses both live near water and find their food there. Both animals likely evolved bills because they are useful for catching and eating food found in the water. The duck and platypus, however, inherited their bills from unrelated ancestors.

LINNAEUS AND A MORE MODERN SYSTEM 33

Duck-billed platypuses are odd creatures. They have a mix of all sorts of different traits.

BY ANY OTHER NAME

Today's scientists are still discovering and naming species. In 2021, scientists named three beetles found in Australia after three rare Pokémon: Articuno, Zapdos, and Moltres. The beetles' names are *Binburrum articuno*, *Binburrum zapdos*, and *Binburrum moltres*. In 2005, the genus of *Acacia* trees and shrubs was split into three different genera by botanists. Australian botanists waged a campaign to allow the Australian species to keep the name *Acacia*. *Acacia* species make up a large percentage of wild and commercially grown plant species in Australia. The botanists successfully argued that changing the scientific names of the almost 1,000 *Acacia* species in Australia would confuse people, hurt the horticulture

One of the many *Acacia* species in Australia is the poverty wattle, shown here.

industry, and require the rewriting of many plant publications. Australian *Acacia* trees, such as *Acacia decora*, were allowed to keep their names. African, Asian, and American acacias now have the new genus names *Vachellia* and *Senegalia*.

When scientists discover a new species or want to rename an old species, the name must meet the requirements of an international organization. Plant names are approved by the International Committee on Systematics of Prokaryotes. New animal species names are approved by the International Commission on Zoological Nomenclature. New bacterial species names are approved by the International Code of Nomenclature of Bacteria.

The requirements for naming species differ from group to group, but the major requirements are the same. Names cannot be repeated. It is also important that both the genus and species names originate in Latin. (As are the kingdom, the phylum, the class, the order, and the family.) People around the world use these scientific names. Latin is not any nation's official language, so it is seen as neutral.

The species name must also follow the binomial naming system used for hundreds of years. Each name must have a genus and species name. The genus is capitalized while the species name is not. Both parts of the name must be italicized. One example is the scientific name of humans: *Homo sapiens*. The genus is *Homo*. The species name is *sapiens*. The two parts together are the overall species name.

Linnaeus's system is one useful way of organizing living things. Except for the rank of species, however, they are somewhat arbitrary. There is no definition for the terms "family" or "class" except to say that one group is bigger than the other. (Families fit within a class.) Though Linnaeus's contributions remain important, other methods of classifying living things would soon arise.

CHAPTER 3

A MATTER OF EVOLUTION

Although many people over the centuries have tried to come up with newer and better ways to classify Earth's plants and animals, perhaps the best known is Charles Darwin (1809–1882). Darwin, an English naturalist, was interested in nature and science from the time he was a child. He studied a variety of things, including medicine, botany, and zoology, before embarking on a voyage that would change everything. His studies and investigations changed the way people around the world view the natural realm.

Earth has many creatures, and there are different ideas about classifying them.

IDEAS OF EVOLUTION

In 1831, Darwin's friend and botany professor John Stevens Henslow recommended that he go on an upcoming ocean voyage with Captain Robert Fitzroy. Fitzroy wanted a naturalist to collect and catalog plants and animals on a trip to South America for the British Royal Navy. Darwin agreed.

The HMS *Beagle* set sail from England in December 1831. The voyagers returned to England almost five years later. During the trip, Darwin wrote detailed notes and drew sketches of the organisms he encountered on his journey. He also collected numerous plants, animal, and mineral specimens. Years later, Darwin used his observations from his *Beagle* journey to come up with his theory of evolution by natural selection.

Charles Darwin originally studied to become a doctor, but he didn't like the classes and often skipped them to wander and learn about things in the wild.

Darwin is often called the father of evolution. However, he was not the first person to suggest that different species descended from common ancestors. Darwin and another scientist, Alfred Russel Wallace, each independently made the major step of explaining evolution and what causes it. Darwin and Wallace pointed out that the organisms that adapted to their surroundings best would be the ones to survive and pass on these useful traits to their offspring. Darwin called this idea natural selection. Darwin's name became forever attached to these ideas in 1859 when he published *On the Origin of Species*.

Members of the same species can breed to produce offspring that also survive and can reproduce. The whole group of a species doesn't have to live in the same place, however. On the *Beagle* trip, Darwin noticed that many similar species were slightly different from one another. He guessed that their differences helped them succeed in their own habitats. For example, on the remote Galápagos Islands, he saw that various species of finches had differently shaped beaks. The islands were too far away from the mainland and too far from each other for finch migration between them. Darwin supposed that the finches on each island probably stayed on that island permanently.

One major difference between the island environments was the types of food available to the finches. On one island, the finches ate insects and had narrow, sharp beaks. On another island, where the finches ate more larvae and buds, the finches' beaks were broad and hooked. There turned out to be 13 individual species of finches that Darwin found. Each had a slightly different beak that was best suited for eating the food most available. Darwin hypothesized that long ago a single type of finch came to live on the islands. Over generations, the finches best equipped for each island survived to pass on their traits to their offspring.

It wasn't just the finches. In England, Darwin observed that domesticated pigeons could be bred to have certain traits. Darwin used his observations on finches, pigeons, and other creatures to come up with an idea of how species arose from common ancestors.

Darwin's process of natural selection has several steps. First, Darwin proposed that offspring with the most useful traits, such as beaks better suited for eating the available seeds, will survive better than other offspring without these traits. Offspring that survive will produce their own offspring. Those offspring will inherit the useful traits from their parents. Different environments lead to different useful traits. Over time, organisms living in diverse environments become distinct enough to develop into different species, and they often lose the ability to interbreed.

The HMS *Beagle* made its voyage in part because Great Britain had a treaty with Argentina. Leaders wanted to know what resources were there.

THE VOYAGE OF THE *BEAGLE*

When the HMS *Beagle* set sail in late December 1831, its voyage was only supposed to last two years. However, it didn't return to England until nearly five years later. Darwin was only 22 when it set out. Throughout the entire voyage, he only wound up spending about 18 months aboard the ship. He spent the rest of the time on land in various locations, studying plants and animals and putting together what would eventually become his theory of evolution by natural selection.

Darwin studied many important things in the Galápagos Islands, but he spent time in other locations during the voyage as well. He visited Argentina and found fossils from many different creatures, including many large land animals. He also studied geology extensively. He saw volcanoes erupt and experienced earthquakes. Darwin trekked up the Andes and found seashells at great heights. He studied atolls and formed more theories. The *Beagle* circumnavigated the world before it returned to England in October 1836.

PHYLOGENETICS

Darwin used the image of a tree to describe some of his ideas. He said species were like the branches of a tree. These branches form when one common ancestor changes over time into many different species. This idea of new species branching off from common ancestors was new in Darwin's time. Now it is an accepted field of study called phylogenetics.

Phylogenetics is the study of evolutionary relationships between organisms. Before Darwin's time, scientists looked for homologous traits to help sort plants and animals into groups. However, they didn't think of these traits as those inherited from common ancestors. Once scientists began using evolutionary relationships, classification made more sense. These relationships could be sketched out in phylogenetic trees.

A phylogenetic tree is similar to a family tree. Family trees have branches that show the relationships between people in a family. In this means of organization, you can easily see the parents of one individual and look back to see other ancestors, such as great-grandparents. Phylogenetic trees show groups instead of individuals. One parent or common ancestral group usually branches off into two groups.

The diagram below is a phylogenetic tree.

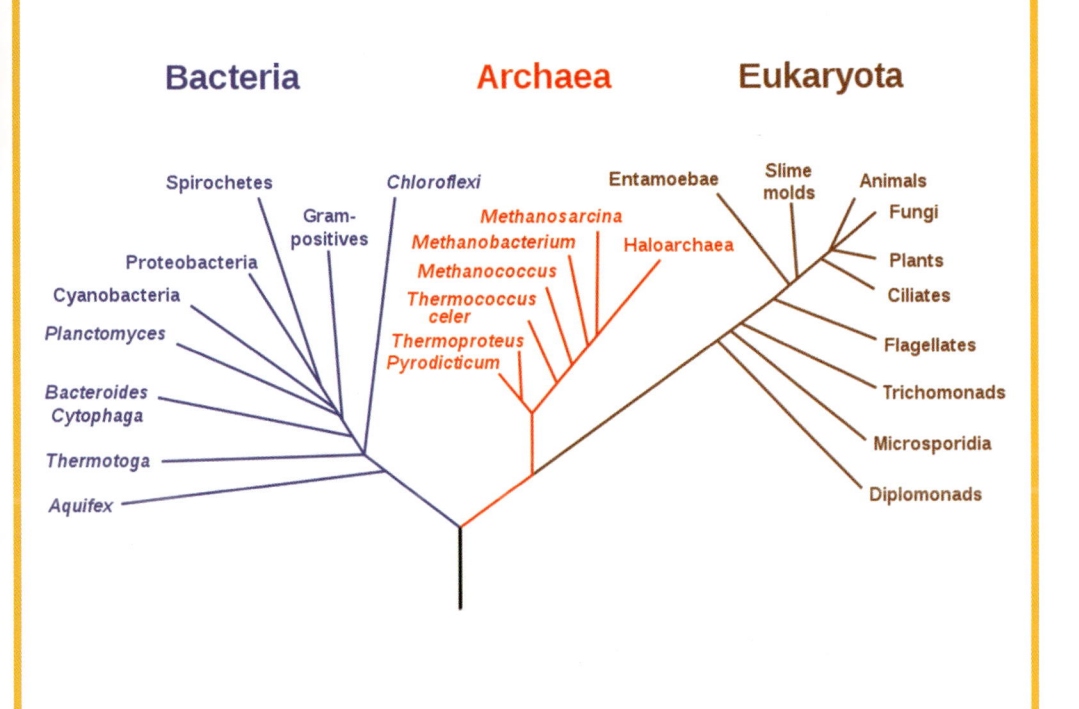

Bacteria
- Spirochetes
- Gram-positives
- Proteobacteria
- Cyanobacteria
- Planctomyces
- Bacteroides Cytophaga
- Thermotoga
- Aquifex

Archaea
- Chloroflexi
- Methanosarcina
- Methanobacterium
- Methanococcus
- Thermococcus celer
- Thermoproteus
- Pyrodicticum
- Entamoebae
- Haloarchaea

Eukaryota
- Slime molds
- Animals
- Fungi
- Plants
- Ciliates
- Flagellates
- Trichomonads
- Microsporidia
- Diplomonads

When scientists correctly classify an organism, they try to find its place in the phylogenetic tree. Correctly classifying organisms can be difficult. The task is easier now, however, than it was before Darwin's theory of evolution by natural selection. Before Darwin, scientists had trouble explaining the ideas that later became known as divergence and convergence. Divergence is when one species splits into others. When one species separates (or becomes separated) into different environments, the groups adapt to become distinct species over time. After Darwin, scientists were able to link differences in similar species to differences in their environments.

Darwin's work helped many different kinds of scientists better study and categorize plants and animals.

A MATTER OF EVOLUTION

Before most scientists accepted the theory of evolution by natural selection, many people didn't understand how or why very distantly related species inherit traits that appear similar. Scientists now know that the similar traits are from different ancestors and have evolved separately; this is called convergence, or parallel evolution. The similar species may look alike, but they are not closely related. They exist in like environments where it is an advantage to have particular traits. Similar traits formed by convergence are called homoplasies.

WHALE SHARKS AND RIGHT WHALES

At first look, whale sharks (*Rhincodon typus*) and North Atlantic right whales (*Eubalaena glacialis*) appear to be closely related. They both are large ocean animals that have large, blunt heads with big mouths. They both have dark gray or brown on the top of their bodies and lighter colors underneath.

Whale sharks and right whales are both filter feeders. They eat similar things. Whale sharks feed by opening their large mouths and sucking in water. Plankton, small fish, small invertebrates, and fish eggs in the water get stuck on the whale shark's gill rakers. These are bristles that catch small creatures in the water before the water enters the whale shark's gills. The plankton and other creatures go into the whale shark's stomach.

Right whales, however, have baleen plates inside their mouths. These plates are similar to gill rakers, but the right whale has baleen plates instead of teeth. The whale shark has gill rakers and teeth. Right whales feed on plankton that gets stuck on their baleen plates.

Though whale sharks live in the warm waters of the world and right whales in colder waters, the ranges of the two species overlap slightly in waters near the

INSIDE BIOLOGICAL TAXONOMY

A MATTER OF EVOLUTION

45

Whale sharks and right whales look similar. However, their traits and descent put them in entirely different taxonomic groups.

Florida coast. Whale sharks and right whales look and act alike because they have adapted to feeding on similar prey in similar environments. However, they come from very different lines of descent. Like other sharks, whale sharks belong to the class Chondrichthyes. The members of this class are cartilaginous fish. Their skeletons are made of cartilage, like that on the tip of the human nose.

Whale sharks are often called the largest fish in the ocean. Fish themselves are not a recognized group in taxonomic classification. However, when people talk about fish, they usually mean the aquatic animals with skeletons, gills, and fins. Whale sharks are not mammals, so by this definition, they are the largest fish in the ocean. They can grow to be up to 40 feet (12.2 m) long or more.

Right whales are bigger than whale sharks. They can grow up to 50 feet (15.24 m) long or longer. Right whales belong to the class Mammalia and the order Cetacea. This order includes all whales, dolphins, and porpoises. Cetaceans are more closely related to cows than to whale sharks.

Whale sharks and right whales have similar features that were passed on from different ancestors and their adaptations. They are a good example of convergence and parallelism. Their large size is useful to both species because few ocean predators will attack animals this huge. Plankton are plentiful in most parts of the ocean, so their diet is also suitable for large animals. Both species come from ancestors that adapted over time to use a sievelike system to get their food.

THE IMPORTANCE OF TRAITS

Every species has different traits, and some of these traits change through evolution and natural selection in quicker ways than others. The traits that change quickly are often ones that are particularly beneficial in different environments. For

A MATTER OF EVOLUTION

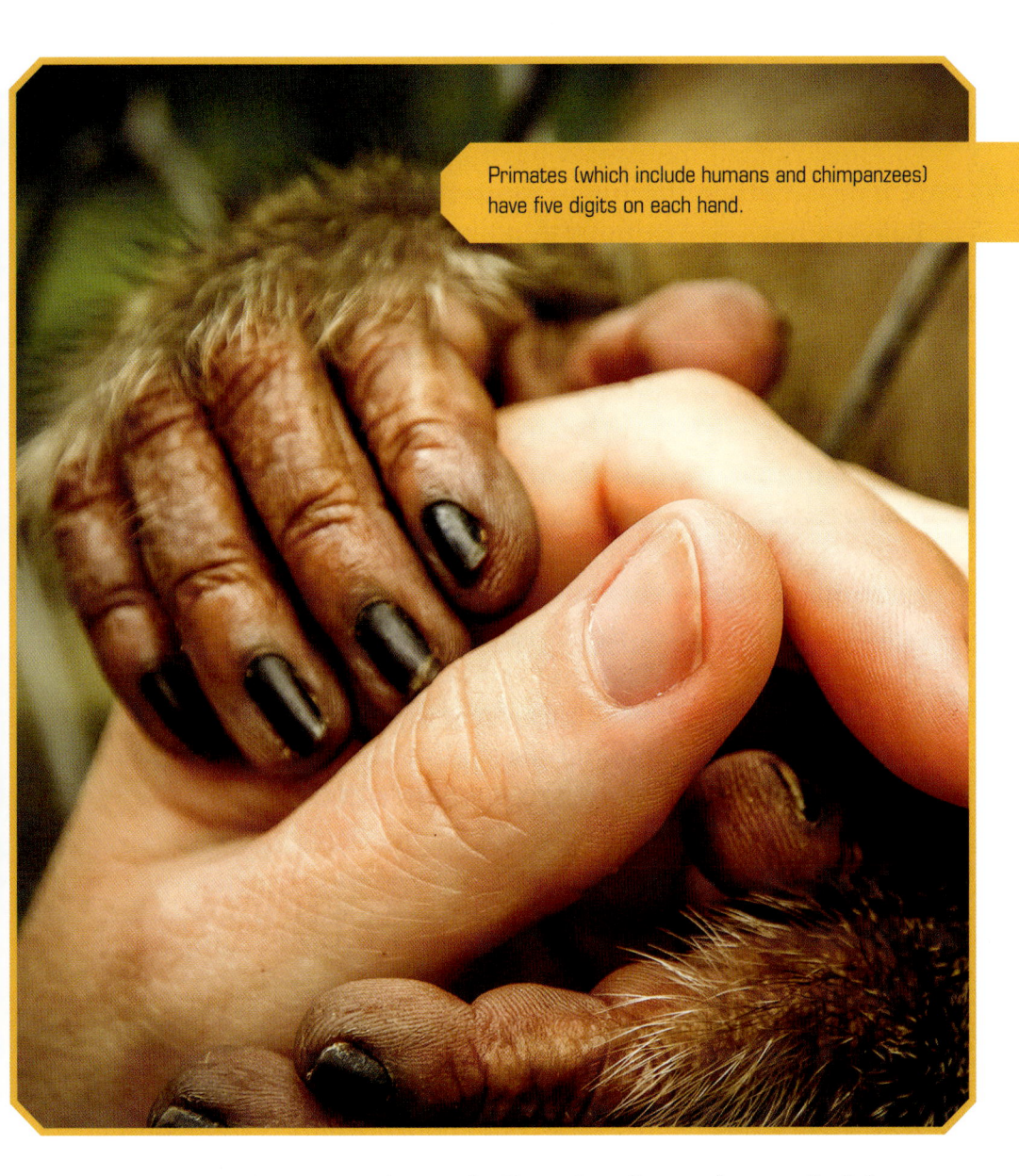

> Primates (which include humans and chimpanzees) have five digits on each hand.

example, all mammals are believed to have descended from an ancestor with five toes or digits. Some mammals, such as primates, still have five digits on their feet. Many, such as horses, do not. Different species of mammals have feet that come in different shapes and have different textures. Foot shape, texture, and the number of digits are characteristics that vary in relation to what is best for a particular mammal species and where it lives.

One trait all mammals have in common is hair, whether it's fur or something else. Beavers have dense fur. Dolphins are born with a tiny amount of whisker-like hair on their chin, though they lose it as they grow. Despite all the differences, all mammals have hair made of a protein called keratin. Modern reptiles, amphibians, fish, and birds do not. This hair on mammals is called a conserved characteristic because it appears on all mammals. Hair on other creatures, such as hairy caterpillars, is not made of keratin. Conserved characteristics do not disappear quickly as species evolve. The fact that these traits last a long time makes these characteristics useful for classifying related animals.

As science advanced after Darwin's discoveries, scientists discovered that many organisms that were classified in the same groups for morphological reasons also shared common traits that could not be obviously seen. For example, plants in the same family may produce similar chemical substances. Some but not all members of the Solanaceae (nightshade) family produce nicotine. Closely related species also share particular DNA sequences. More ways to analyze species can make it easier to see how those species are related. New information can also change ideas about classifications. Scientists are still changing or deciding many classifications based on new evidence.

A MATTER OF EVOLUTION

A dolphin may not seem like it has hair or fur, but it's born with some and fits in with other mammals.

CHAPTER 4

CLADES, CLADISTS, AND CLADISTICS

There have been many trials over the years for scientists who work to classify living organisms. There were struggles over how to classify living organisms using phylogenetics, or the evolutionary relationships each organism had to other organisms. One way to figure out how and where different species branch off from one another in phylogenetic trees is to use something called cladistic analysis, or cladistics.

A clade is a group made of an ancestor and all its descendants. The Greek word *clados* means "branch." Cladists are people who use cladistics. Looking at shared traits, they investigate branching relationships within a selected group of species. Some things they might analyze are an animal's beak shape, tooth length, and brain size. A cladist uses branching relationships to make logical assumptions about which traits a common ancestor had. The common ancestor's form of the trait is called a conservative trait or a plesiomorphy. The changed form of a trait that some newer ancestors have adapted is called a derived trait or an apomorphy. The branched diagram created by this analysis is called a cladogram. Cladistic analysis can use physical, molecular, or behavioral characteristics. Today, scientists often use DNA or RNA base sequences. When scientists use DNA to construct a cladogram, the shared characters they use are small sections of DNA. They look for sections of the gene code that remain stable and others that change in related organisms. The organisms with the most sections that are the same are generally thought to be the most closely related.

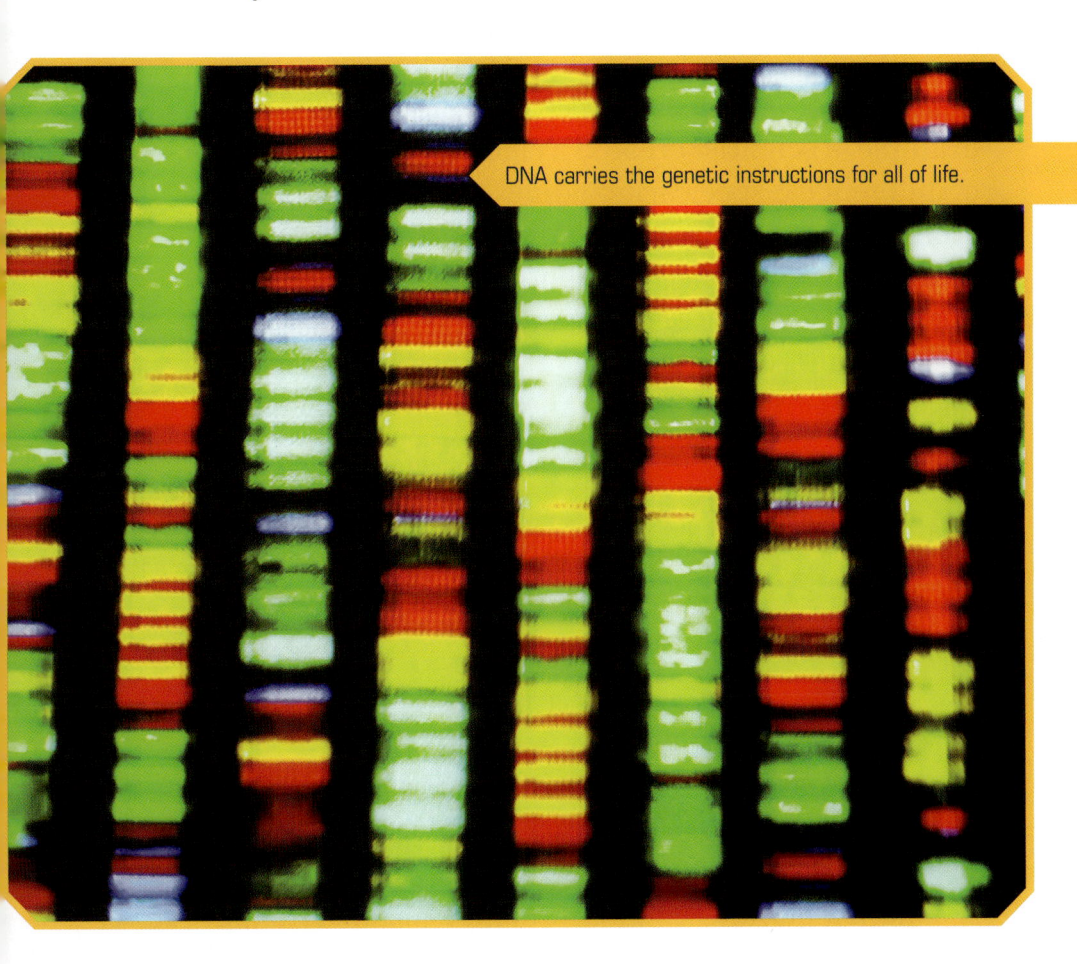

DNA carries the genetic instructions for all of life.

When a clade branches, scientists often assume that it divides into two new lineages. If there seem to be multiple splits, they note that there needs to be further study. Like all classification techniques, cladistic analysis can be tricky. For example, one species may share an apomorphy, or specialized trait, with one species and a different apomorphy with another. The study of cladistics looks for the simplest explanation when investigating different results. Using the simplest explanation in cladistics is called parsimony.

A German entomologist named Willi Hennig was the first to make the use of cladistic analysis popular. Hennig said that cladistics should only be used to infer relationships between the species being analyzed. For example, a scientist created a cladogram using traits of different cat species, including domestic cats, leopards, and lions. The cladogram could only be used to see how those cat species might be related. According to Hennig, that cladogram could not be used to guess how the different cat species are related to domestic dogs because dogs were not included in the analysis.

Hennig also suggested that only synapomorphies can be used to decide which species are most closely related to one another. Synapomorphies are apomorphies (new traits that are different from the trait of a common ancestor) that are shared between species. Cladists theorize that if a new form of a trait is found in two different species, it is more likely that they inherited the new trait from the same ancestor than adapted to develop the same trait twice.

Cladograms represent nested phylogenies. "Nested" means that the bigger group of the clade is subdivided into smaller related sister groups, which are groups that branch off from each other on the cladogram.

A cladogram that shows the order in which different related species branched off from each other is called a rooted cladogram. A cladogram that shows relationships with no chronological order is said to be unrooted.

CLADES, CLADISTS, AND CLADISTICS 53

A cladogram of cat species could be used to see how lions and tigers are related. However, it couldn't be used to see how lions and wolves are related.

CLADISTICS CLASHES

Cladistics can sometimes disagree with the classification system developed by traditional taxonomists. For example, for a long time, all birds, including robins, blue jays, and pink flamingos, have belonged to a class called Aves. All snakes, lizards, crocodiles, and other scaly land animals have belonged in a class called Reptilia. Dogs, bears, and people belong to a class called Mammalia. Cladists, however, disagree with this way of classification. If representative species from each group are analyzed on a rooted cladogram, the cladogram would show that all three groups share a common ancestor. A saltwater crocodile, however, would be evolutionarily closer to this ancestor than a pink flamingo because the crocodile and the common ancestor share lots of characteristics. A German shepherd, which shares the fewest common characteristics, would be evolutionarily the furthest away from the common ancestor.

HENNIG AND CLADISTICS

Willi Hennig (1913–1976) called his method of cladistic analysis "phylogenetic systematics." He never actually used the term "cladistics" in his writings, which focused on the relationships among the larvae of Diptera (the order of insects that includes flies, mosquitoes, and gnats). The word "cladistic" was first used by other zoologists working in the 1960s. In his writings, Hennig did identify an ancestor species and its entire descendant species as a clade. He was not the first person to suggest using the methods he described, but he was the first person to give clear rules.

Humans are often classified in their own family, Hominidae. Other primates, such as chimpanzees, apes, and spider monkeys, are placed in an entirely different family called Pongidae. If scientists create a cladogram by analyzing DNA, they can see that chimpanzees and humans are more closely related than chimpanzees and spider monkeys. Primatologists, taxonomists, and cladists are still figuring out how the primate order should be divided.

Chimpanzees are primates just like humans. However, they're in a different family.

CHAPTER 5

THE FUTURE OF CLASSIFICATION

The classification of living creatures has occupied people for a long time, back to the time of ancient Greece and beyond. It's changed a lot over that time, and it will continue to change into the future.

For more than 2,000 years, scientists classified living creatures into two groups: plants and animals. The differences seem clear. Plants are often green and grow from the ground. Animals are often very visible and noisy, and they can move about. Since the invention of microscopes in the 1600s, scientists have learned even more about the differences between the two. They go right down to the microscopic level. Plant cells have a thick wall around them. Animal cells have a thin membrane. Scientists called these large groups "kingdoms." While the idea of kingdoms is still very important, the science of classification has changed a lot over the years.

Dutch naturalist Antonie van Leeuwenhoek (1632–1723) discovered animalcules, or microscopic organisms—some in his own bodily waste. They were very simple creatures. Still, some photosynthesized like plants, and some swam around, powered by tiny tails called flagella. Scientists used these observations to classify them as plants or animals. Other microscopic organisms, however, foiled scientists' efforts. The euglena, a single-celled creature discovered by Leeuwenhoek, was mobile and had a primitive eyespot, but it also had the ability to photosynthesize. So, what was it?

MORE KINGDOMS

These issues suggested the need for more kingdoms. In 1866, a German scientist named Ernst Haeckel suggested a third kingdom called Protista.

Haeckel was a firm believer in Darwin's theory of evolution. He believed that very similar organisms, such as dogs and wolves, came from a common ancestor relatively recently. He reasoned, however, that very different organisms, like dogs and houseflies, had a common ancestor much longer ago. The common ancestor of dogs and common bacteria would have existed even longer ago still.

It's pretty obvious that a dog and a wolf (like the gray wolf shown) had a common ancestor not so long ago. But what about dogs and bacteria?

Haeckel hypothesized that the new microscopic organisms that Leeuwenhoek had described might be the common ancestor of all plants and animals. These microorganisms were so small, so simple, and so completely different from other animals and plants that they could be the most ancient common ancestors of all life.

Haeckel noticed that the microorganisms were similar to both plants and animals in some ways. If these creatures were the common ancestors of all organisms, he pointed out, then they could not be called a plant or an animal. So Haeckel suggested that these tiny living creatures should be placed into their own kingdom.

This creation of a third kingdom signaled a tremendous change in classification. People were used to just two kingdoms, and until Haeckel's time, scientists grouped life based solely on an organism's appearance or behavior. Tigers, lions, and cheetahs all have a furry, sleek appearance and are carnivorous, meaning that they eat other animals. They are easily grouped as animals, particularly as cats. Fir trees, hemlock trees, and blue spruce trees all have needles and cones. They are easily classified as pines. These living organisms are easily grouped by their appearance and sometimes behavior.

Darwin's theory of evolution by natural selection showed that life began with simple organisms and changed over a very long time. Haeckel used this idea to group creatures according to how they were related. It helped when microscopes became more powerful through the early part of the 20th century. Scientists studying microorganisms began to realize just how different they were from each other. Perhaps, some people thought, they should not all be in the same kingdom.

In 1956, a scientist named Herbert Copeland tried to solve this problem by splitting Haeckel's kingdom Protista into

THE FUTURE OF CLASSIFICATION

> Ernst Haeckel discovered, named, and described thousands of new species. He also created many new terms in the field of biology.

Dinoflagellates like these are single-celled eukaryotes.

two groups. He put the prokaryotes, which had no nucleus, in a new kingdom he called Mychota. He put the eukaryotes, with their complex cell structure, in a group he called Protoctista. This group contained red algae, dinoflagellates, fungi, and single-celled creatures that had a nucleus.

Copeland was just one of a large group of scientists who realized that having a nucleus and complex cell structure were enormous steps in evolution. Bacteria clearly were quite simple. They had only a cell wall, cytoplasm (everything contained within the plasma membrane cell), and sometimes one or more whiplike structures called flagella. These served to move them around. Bacteria have fewer specialized structures within the cell: only photosynthetic membrane systems and gas vesicles.

Eukaryotes have a nucleus that holds their DNA. They also have special structures to help make food, digest food, manufacture proteins, and serve many other roles. Scientists saw that these structures function like the organs in an animal and called them organelles.

It was not long before scientists saw that another group of organisms was different enough to deserve its own kingdom. In the past, fungi had been included in the plant kingdom or (in Copeland's system) Protoctista. Like plants, fungi grow out of the ground or on other materials, and they do not move around as animals do.

INSIDE BIOLOGICAL TAXONOMY
62

Mushrooms, such as the ones shown in this photo, are fungi. They may seem a lot like plants, but there are many key differences.

In 1959, American biologist Robert H. Whittaker (1920–1980) proposed that fungi are quite different from plants and should have their own kingdom. Whittaker pointed out that fungi hold a very different place in the food chain. Plants are called producers. They use chlorophyll to trap the energy of sunlight and produce sugar. Trees, bushes, and grasses all use green leaves to capture this energy, make food, and grow. Plants are then eaten by many other life-forms. Nearly all of Earth's energy comes from the sun, and this forms the base of the food chain.

Fungi, however, hold a completely different place on the food chain. Fungi break down, or digest, other organisms for their food. Most people first see fungi on rotting food. When food goes bad or gets moldy, it is usually a fungus growing on the food, digesting it. The mold absorbs the nutrients and uses them for energy.

By recognizing Fungi as a separate kingdom, Whittaker suggested that there were five kingdoms: Plantae, Animalia, Monera, Protista, and Fungi. Many people quickly accepted this arrangement of five kingdoms, which is still used in many places today. Changes to the tree of life did not stop, however.

NEW TECHNOLOGY, NEW TOOLS

Years ago, the invention of the microscope led to new discoveries. In somewhat more recent years, new technologies again played a role in a major advance. By the early 1970s, scientists could sequence, or read the order of, DNA and RNA. These genetic molecules make up the instructions that an organism uses to live, grow, and reproduce. Parts of DNA are called genes. Some of these genes can be found in all organisms. These genes change bit by bit as populations change. Close relatives are very similar genetically. More distant relatives are much less similar genetically.

In the early 1970s, Carl R. Woese, an American molecular biologist, was studying bacterial RNA. Woese and his team compared one particular gene that was found in a mouse, in duckweed, in yeast, and in a few species of bacteria. The results proved to be historic.

Woese and his team confirmed that the mouse, the duckweed, and the yeast were closely related—surprisingly close, considering how different the organisms look. The big surprise, however, came when Woese examined the relationships between the bacteria. Two of the species of bacteria, called methanogens because they produce methane, showed that they were very distant relatives of the other two species of bacteria. In fact, the methane producers were so distantly related to the other bacteria that they were actually closer relatives to the mouse, the duckweed, and the yeast. It was almost as if they should be in their own kingdom.

THE FUTURE OF CLASSIFICATION

Black smokers can have liquid water that's far above water's usual boiling point.

EXTREME!

Many archaea (a kind of single-celled organism) and some bacteria live in particularly extreme living conditions. Because of this, they're called extremophiles. For many years, people thought that boiling water and cooking food killed all microorganisms. Cooking is important to the safety of food. However, scientists have found members of the domain Archaea living in the hot springs of Yellowstone National Park, where water temperatures average near boiling! These organisms, called hyperthermophiles, don't just survive in these temperatures, they thrive. In many cases, these prokaryotes reproduce faster as the temperature increases.

Scientists have found some extremophiles living near black smokers at the bottom of the Pacific Ocean, at depths of 7,000 feet (2,134 m). A black smoker is a vent, or hole, from which superheated mineral-rich water rises from deep within Earth's crust. One extremophile is a type of bacteria that lives within the giant tube worms that live near these smokers. These waters may reach temperatures of 750°F (399°C)! That is well above the boiling point of water, but this water does not boil because of the weight of the ocean's water at these depths. Another type of extremophile called an acidophile thrives in incredibly acidic and sulfur-rich fields near active volcanoes in Japan and Italy. The ability of these life-forms to withstand such conditions has led some scientists to speculate that these forms of life may have been carried to Earth from other planets by meteorites. No one is sure, however.

Woese divided different types of bacteria into two groups, Archaea and Bacteria. He did not make these groups kingdoms. The two groups were such distant relatives that Woese created a whole new rank above kingdom called a domain.

THE FUTURE OF CLASSIFICATION

The domain Bacteria contains many single-celled organisms. The Bacteria species *Escherichia coli*, popularly known as *E. coli*, lives in people's bodies and helps digestion—but sometimes it can make them sick. The species *Mycobacterium tuberculosis* causes the deadly disease tuberculosis.

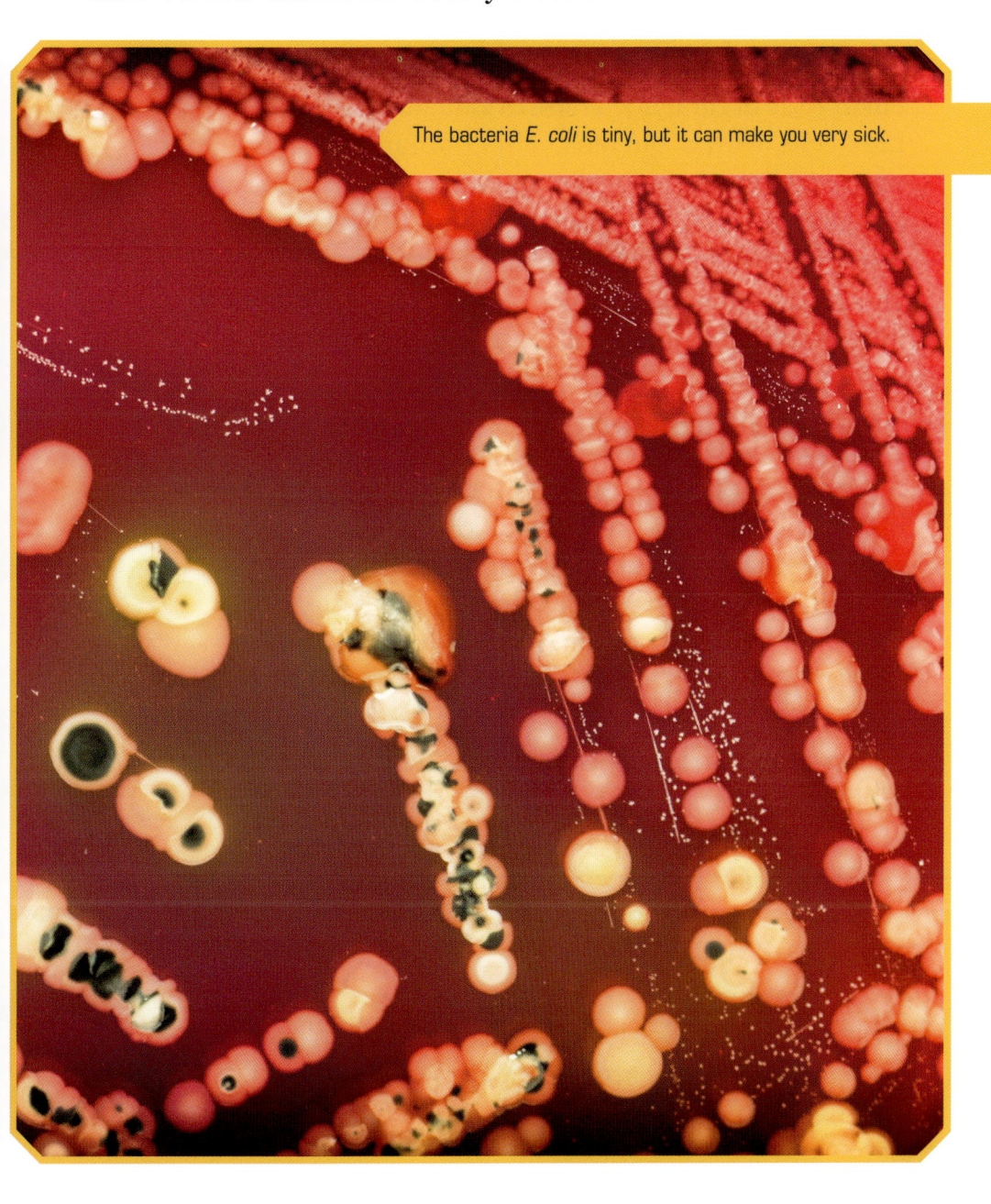

The bacteria *E. coli* is tiny, but it can make you very sick.

The domain Archaea is also made up of single-celled organisms. The members of Archaea differ from those of Bacteria in the proteins within their genetic material. They are found throughout the world, in the oceans as well as in environments such as hot springs.

The third domain, Eukarya, is the domain of eukaryotic organisms. These organisms have complex cells with a nucleus. This domain includes plants, animals, and fungi, as well as protists. All the kingdoms that were once thought to exist are really only a small part of the living world. The species of Archaea and Bacteria easily outweigh all the animals, plants, and fungi on this planet!

ALWAYS CHANGING

Science keeps changing. The names and classifications of organisms are based on relationships to other organisms, and people are never absolutely certain what those relationships are. Scientists try their best to figure out the relationships using fossils, DNA, and other methods, but they can always find new evidence. With new evidence, there will be new ways to classify organisms. This means the system of classification will always be changing, and the tree of life will never be truly complete.

In the past, the best systems of classification considered animals and plants the only organisms. Fungi used to be included in the plant kingdom. Scientists now know that fungi are more closely related to animals! Scientists also know that plants, animals, and fungi are only a tiny fraction of all life on Earth. There is so much left to find.

THE FUTURE OF CLASSIFICATION

69

Scientists are learning more about life on Earth every day. There's a lot to know and learn!

GLOSSARY

APOMORPHY A trait that is unique to an ancestral species and all descendants.

CLADISTICS The classification of organisms into an evolutionary tree based solely on shared synapomorphies.

CLASS The category in the Linnaean system of classification just below phylum.

CONSERVED CHARACTERISTIC A trait that does not disappear quickly as a species evolves; this is useful in finding relationships between organisms.

DOMAIN The highest taxonomic rank. There are three domains: Archaea, Bacteria, and Eukarya.

DORSAL NERVE CORD A hollow cord that runs dorsal (along the back) to the notochord. It is present for part of the life cycle in all of the phylum Chordata.

EUKARYOTE An organism whose cell or cells contain nuclei and other complex structures with membranes.

FAMILY The category in the Linnaean system of classification just below order.

GENUS The category in the Linnaean system of classification just below family; it is a mandatory part of a species name.

HOMOLOGY Any trait shared between organisms that is due to shared ancestry.

HOMOPLASY Similar traits formed by convergence.

INFER To reach a conclusion after having made a series of observations.

KINGDOM Highest category in the Linnaean system of classification. There were originally two kingdoms: plants and animals. One current system recognizes six.

GLOSSARY

MICROORGANISM An organism that is invisible to the naked eye, thus microscopic, and usually single-celled.

MORPHOLOGY The structure and form of an organism; it is often used to classify the organism.

NOTOCHORD A flexible support structure to which muscles may attach. It is present for part of the life cycle in all of the phylum Chordata.

ORDER The category in the Linnaean system of classification just below class.

PARSIMONY The assumption that traits of an organism are unlikely to evolve twice and so are more likely to be synapomorphies.

PHARYNGEAL SLIT A perforation in the pharynx wall used to filter food; these are present for part of the life cycle in all of the phyla Chordata and Hemichordata.

PHYLOGENETICS Using evolutionary relationships to classify organisms.

PHYLUM The category in the Linnaean system of classification just below kingdom.

PROKARYOTE A single-celled microorganism classified as Archaea or Bacteria that has a very simple cell structure and no nuclei or other membrane-bound organelles.

SPECIES A name that differentiates the organism from all others in the genus; it is part of Aristotle's naming system. Every species is identified by a binomial consisting of its genus and species.

SYNAPOMORPHY A trait shared among groups that came from their common ancestor.

SYSTEMATICS The study of the diversity of life and the relationships between organisms.

TAXONOMY The science of classifying living organisms.

FOR MORE INFORMATION

AMERICAN MUSEUM OF NATURAL HISTORY
200 Central Park West
New York, NY 10024-5102
(212) 769-5100
Website: www.amnh.org
This museum researches and interprets its collections and helps teach the public about various cultures and the natural world.

AMERICAN SOCIETY OF PLANT TAXONOMISTS
4344 Shaw Boulevard
St. Louis, MO 63110
Website: www.aspt.net
This group promotes the research and teaching of taxonomy, systematics, and phylogeny of plants.

LINNEAN SOCIETY OF LONDON
Burlington House
Piccadilly, London W1J 0BF
England
Website: linnean.org
The Linnean Society of London was founded in 1788 and is the world's oldest active biological society. The organization works to cultivate the science of natural history.

NATIONAL MUSEUM OF NATURAL HISTORY
Smithsonian Institution
10th Street and Constitution Avenue NW
Washington, DC 20560
Website: naturalhistory.si.edu
The National Museum of Natural History provides research, exhibitions, collections, and education programs on natural science.

FOR MORE INFORMATION

NATIONAL SCIENCE FOUNDATION
2415 Eisenhower Avenue
Arlington, VA 22314
(703) 292-5111
Website: www.nsf.gov
This agency was formed by the U.S. government "to promote the progress of science; to advance the national health, prosperity, and welfare; to secure the national defense."

UNESCO
7 Place de Fontenoy
75352 Paris 075P
France
Website: www.unesco.org
UNESCO promotes international cooperation in the sciences and carries out freshwater, marine, ecological, and earth science programs.

U.S. FISH AND WILDLIFE SERVICE
Division of Information Resources and Technology Management
5275 Leesburg Pike
Falls Church, VA 22041
(703) 358-1729
Website: www.fws.gov
A bureau within the Department of the Interior, the U.S. Fish and Wildlife Service works to preserve, protect, and help the habitats of fish, wildlife, and plants.

WILLI HENNIG SOCIETY
Invertebrate Zoology
American Museum of Natural History
PO Box 37012
New York, NY 20013-7012
(202) 633-1005
Website: www.cladistics.org
The Willi Hennig Society promotes the field of phylogenetic systematics.

FOR FURTHER READING

Angier, Natalie. *The Canon: A Whirligig Tour of the Beautiful Basics of Science.* New York, NY: Houghton Mifflin Harcourt, 2017.

Bainbridge, David. *How Zoologists Organize Things: The Art of Classification.* London, UK: White Lion Publishing, 2020.

Beil, Karen Magnuson. *What Linnaeus Saw: A Scientist's Quest to Name Every Living Thing.* New York, NY: Norton Young Readers, 2019.

Broom, Jenny. *Animalium: Welcome to the Museum.* Dorking, UK: Big Picture Press, 2014.

Claybourne, Anna. *Amazing Evolution: The Journey of Life.* New York, NY: Ivy Kids, 2019.

Crowder, Bland. *National Geographic Pocket Guide to Trees and Shrubs of North America.* Washington, DC: National Geographic Society, 2015.

Dunn, Rob. *Every Living Thing: Man's Obsessive Quest to Catalog Life, from Nanobacteria to New Monkeys.* New York, NY: HarperCollins, 2010.

Nijhuis, Michelle. *Beloved Beasts: Fighting for Life in an Age of Extinction.* New York, NY: W. W. Norton & Company, Inc, 2021.

Pobst, Sandra. *National Geographic Investigates: Animals on the Edge: Science Races to Save Species Threatened with Extinction.* Des Moines, IA: National Geographic Children's Books, 2008.

FOR FURTHER READING

Royston, Angela. *Animal Classification*. New York, NY: Gareth Stevens, 2013.

Sartore, Joel. *The Photo Ark: One Man's Quest to Document the World's Animals*. Washington, DC: National Geographic, 2020.

Simpson, Kathleen. *National Geographic Investigates: Genetics: From DNA to Designer Dogs*. Des Moines, IA: National Geographic Children's Books, 2008.

Strager, Hanne. *A Modest Genius: The Story of Darwin's Life and How His Ideas Changed Everything*. CreateSpace Independent Publishing Platform, 2016.

Strauss, Rochelle. *Tree of Life: The Incredible Biodiversity of Life on Earth*. Toronto, ON: Kids Can Press, 2013.

Winston, Robert. *Evolution Revolution*. New York, NY: DK Publishing, 2016.

INDEX

A
Acacia, splitting of genus, 34–35
Acer saccharum, 24, 28
acidophiles, 66
animalcules, 56
Animalia, 28, 29, 63
apomorphies, 50, 51, 52
appearance, as basis for classification, 58
Archaea, 66, 68
Aristotle, 7, 8, 10, 11, 12, 13, 14, 16, 17, 27
Aves, 28, 31, 54

B
bacteria, 35, 57, 61, 64, 66, 67
Bacteria (domain), 66, 67, 68
Bauhin, Gaspard and Jean, 18, 22
behavior, as basis for classification, 8, 58
Binburrum articuno, 34
Binburrum moltres, 34
Binburrum zapdos, 34
binomial definition, 11
binomial nomenclature, 22, 24, 35
biological taxonomy, definition of, 5

C
Catalogue of Cambridge Plants, 18
Cesalpino, Andrea, 17, 18
Cetacea, 46
Chondrichthyes, 46
Chordata, 28, 31
clades, 50, 51, 52, 54
cladistic analysis/cladistics, 50–52, 54–55
cladists, 50, 52, 54, 55
cladograms, 50, 52, 53, 54, 55
class, as classification group, 27, 28, 35
color, as basis for classification, 8
conservative trait, 50
conserved characteristic, 48
convergence, 42, 43, 46
Copeland, Herbert, 58–61
cotyledons, as basis for classification, 18–20

INDEX

D
Darwin, Charles, 36, 37–39, 40, 42, 48, 57, 58
derived trait, 50
dichotomous key, 28–29
dicots, 20
Diptera, 54
divergence, 42
domains, 66, 67, 68
duck and duck-billed platypus, comparison of, 32
duck-billed platypus, 32, 33

E
embryology, 31–32
embryonic development, as basis for classification, 31–32
Eudyptes, 28, 31
euglena, 57
Eukarya, 68
eukaryotes, 61, 68
evolution/evolution by natural selection, 12, 13, 37–39, 40, 42, 43, 46, 57, 58, 61
evolutionary relationships, as basis for classification, 41
extremophiles, 66

F
family, as classification group, 27, 28, 35
Fitzroy, Captain Robert, 37
fungi, 61–63, 68
Fungi (kingdom), 63

G
genus, as classification group, 11, 18, 25, 27, 28

H
habitat, as basis for classification, 8
Haeckel, Ernst, 57, 58, 59
hair/fur, 32, 48, 49
Hennig, Willi, 52, 54
Henslow, John Stevens, 37
HMS *Beagle*, 37, 38, 39, 40
Hominidae, 55
homologous characters/traits, 32, 40
homology, 32
homoplastic traits/homoplasies, 32, 43
Homo sapiens, 35
hyperthermophiles, 66

I
International Code of Nomenclature of Bacteria, 35
International Commission on Zoological Nomenclature, 35
International Committee on Systematics of Prokaryotes, 35

K

killer whales and great white sharks, comparison of, 8, 9, 10–11
kingdom, as classification group, 27, 28, 56, 57–63

L

Leeuwenhoek, Antonie van, 56, 57, 58
Lepidoptera, 28, 31
Linnaeus, Carolus, 21–24, 26, 27, 28, 31, 35
lions, 25, 52, 53, 58

M

Mammalia/mammals, 10, 12, 18, 20, 32, 46, 47–48, 49, 54
methanogens, 64
microscopes, 56, 58, 64
microscopic organisms/microorganisms, 56, 58, 66
Monera, 63
monocots, 20
morphology, as basis for classification, 18, 28, 29, 31, 48
Mychota, 61

N

nested phylogenies, 52
Nymphalidae, 28, 31

O

On the Origin of Species, 38
order, as classification group, 27, 28

P

parallel evolution, 43, 46
parsimony, 52
phylogenetics, 40–43, 50
phylogenetic systematics, 54
phylogenetic tree, 41, 42, 50
phylum, as classification group, 27, 28
Plantae, 28, 29, 63
plesiomorphy, 50
Pongidae, 55
Primata, 27
primates, 27, 47, 55
prokaryotes, 61, 66
Protista, 57, 58, 63
Protoctista, 61

R

Ray, John, 18–20, 28
Renaissance, 16
Reptilia, 54

rockhopper penguins and painted lady butterflies, comparison of, 29–31
rooted cladogram, 52, 54

S

scientific names, 24, 25, 35
Senegalia, 35
Siegesbeck, Johann, 23
species, as classification group, 11, 24, 27, 28, 35
species, naming requirements for, 35
Spheniscidae, 28, 31
Sphenisciformes, 28, 31
synapomorphies, 52
Systema Naturae, 26, 27, 28

T

taxonomy, derivation of word, 6
Theophrastus, 14
tigers, 25, 53, 58
Tillandz, Elias, 23

U

unrooted cladogram, 52

U

Vachellia, 35
Vanessa, 28, 31
Vesalius, Andreas, 17

W

Wallace, Alfred Russel, 38
whale sharks and right whales, comparison of, 43–46
Whittaker, Robert H., 63
Woese, Carl R., 64, 66

PHOTO CREDITS

Cover, pp. 1–80 Who is Danny/Shutterstock.com; cover Dennis Sadlowski/Shutterstock.com; p. 4 Motortion Films/Shutterstock.com; p. 5 Yobab/Shutterstock.com; p. 6 Mameraman/Shutterstock.com; p. 7 https://commons.wikimedia.org/wiki/File:Rembrandt_-_Aristotle_with_a_Bust_of_Homer_-_WGA19232.jpg; p. 9 Andrea Izzotti/Shutterstock.com; p. 9 hanhanpeggy/iStock/Getty Images; p. 11 Alexius Sutandio/Shutterstock.com; p. 12 Elena Arkadova/Shutterstock.com; p. 15 Scott Shymko/Moment/Getty Images; p. 17 https://commons.wikimedia.org/wiki/File:Andrea_Cesalpino.jpg; p. 19 https://commons.wikimedia.org/wiki/File:John_Ray_from_NPG.jpg; p. 20 AlessandroZocc/Shutterstock.com; p. 21 https://commons.wikimedia.org/wiki/File:LA2-Rashult-2.jpg; p. 23 Time Life Pictures/Contributor/The LIFE Picture Collection/Getty Images; p. 24 JHVEPhoto/Shutterstock.com; p. 25 Maggy Meyer/Shutterstock.com; p. 26 https://commons.wikimedia.org/wiki/File:Linnaeus1758-title-page.jpg; p. 30 Sean Xu/Shutterstock.com; p. 30 jo Crebbin/Shutterstock.com; p. 33 Joao Inacio/Moment/Getty Images; p. 34 Lev Kropotov/Shutterstock.com; p. 36 PopTika/Shutterstock.com; p. 37 https://commons.wikimedia.org/wiki/File:Charles_Darwin_by_G._Richmond.jpg; p. 39 DEA Picture Library/Contributor/De Agostini/Getty Images; p. 41 https://commons.wikimedia.org/wiki/File:Phylogenetic_tree.svg; p. 42 Elnur/Shutterstock.com; p. 44 orifec_a31/Shutterstock.com; p. 45 Foto 4440/Shutterstock.com; p. 47 Kostya Zatulin/Shutterstock.com; p. 49 Ricardo Canino/Shutterstock.com; p. 51 Gio.tto/Shutterstock.com; p. 53 Eleanor Esterhuizen/Shutterstock.com; p. 55 Bildagentur Zoonar GmbH/Shutterstock.com; p. 57 Holly Kuchera/Shutterstock.com; p. 59 https://commons.wikimedia.org/wiki/File:ErnstHaeckel.jpg; p. 60 Roland Birke/Photodisc/Getty Images; p. 62 Tienuskin/Shutterstock.com; p. 65 https://commons.wikimedia.org/wiki/File:Champagne_vent_white_smokers.jpg; p. 67 AnaLysiSStudiO/Shutterstock.com; p. 69 Gorodenkoff/Shutterstock.com.